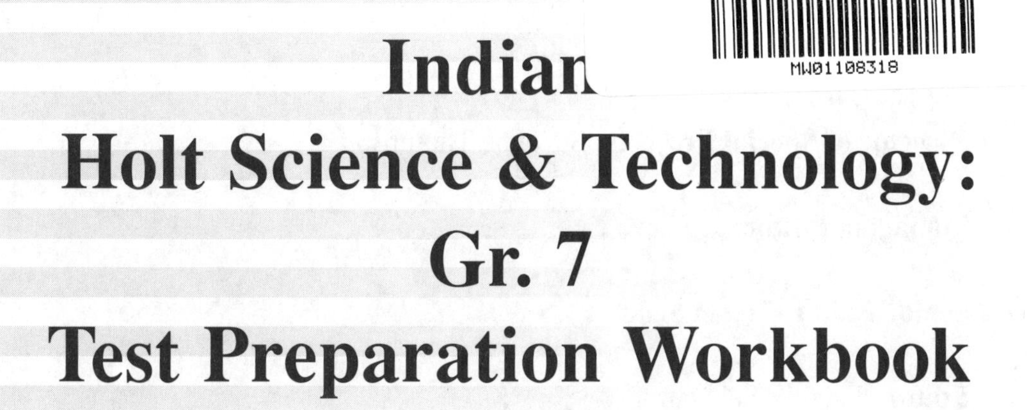

Indian
Holt Science & Technology: Gr. 7
Test Preparation Workbook

HOLT, RINEHART AND WINSTON

A Harcourt Education Company

Orlando • Austin • New York • San Diego • Toronto • London

STAFF CREDITS

Director of Special Projects: Suzanne Thompson

Managing Editor: Jenvieve Eyre

Senior Editor: Tressa Sanders

Editor: Jennifer Schwan

Printed in the United States of America

ISBN 0-03-039897-5

1 2 3 4 5 6 7 8 9 032 08 07 06 05 04

Contents

Standardized Test Preparation Questions

Preparing Students for the Indiana Statewide Testing for Educational Progress

Introduction

This practice book consists of multiple-choice and constructed response questions, representative of the questions on the ISTEP+ Assessment. While this book prepares students for the ISTEP+, no questions contained within are actual ISTEP+ items, nor is this preparation book exactly like the ISTEP+ Assessment.

These practice questions should be used as a part of the local assessment program. A student's performance on the practice questions can serve as an indication of those areas that need additional review and practice. The breadth of content coverage included in this booklet provides teachers an opportunity to assess their students' strengths and weaknesses, as well as the strengths and weaknesses of the overall science program.

Because each page is correlated to a specific chapter or unit of *Indiana Grade 7 Holt Science and Technology,* finding support materials for students to use for review and reinforcement is an easy task. By referring to the relevant section of the textbook, a student can review fundamental science vocabulary, concepts, and principles, and use the questions at the end of the section or chapter for additional review and testing.

Name________________________ Date__________ Class______________

Chapter 1 Standardized Test Preparation

1. A possible explanation or answer to a question is called a(n)

   A. conclusion.

   B. hypothesis.

   C. observation.

   D. inference.

2. Pieces of information acquired through experimentation are called

   A. observations.

   B. data.

   C. conclusions.

   D. variables.

3. In science, a unifying explanation for a broad range of hypotheses and observations that have been supported by testing is called a(n)

   A. theory.

   B. educated guess.

   C. inference.

   D. conclusion.

4. A representation of an object or system that helps scientists visualize and understand information is called a

   A. theory.

   B. law.

   C. model.

   D. hypothesis.

Name____________________________ Date __________ Class ______________

Chapter 1 Standardized Test Preparation

5. If research groups studying the benefits of a new medicine report results that are quite different, what can you conclude?

   A. You should accept the results of the researchers at the larger medical center.

   B. You should assume both set of results are not accurate.

   C. More research is needed to determine what caused the results to be different.

   D. One of the groups did not do the research correctly.

6. A sheet of paper 4 cm wide and 5 cm long has an area of

   A. 20 cm.

   B. 9 cm.

   C. 20 cm^2.

   D. 9 cm^2.

7. The density of a metal bar that has a mass of 150 g and a volume of 40 cm^3 is

   A. 3.75 g.

   B. 190 g/cm^3.

   C. 190 g.

   D. 3.75 g/cm^3.

8. The standard system of measurement used by scientists around the world is called the

   A. Global Measurement System.

   B. International System of Units, or SI.

   C. Scientific System of Measurement.

   D. Unified Measurement System.

Name_____________________________ Date__________ Class_______________

Chapter 1 Standardized Test Preparation

Answer the following questions on a separate piece of paper.

1. A commercial on television tells you that "four out of five dentists who used this product recommended it." Why is this not enough evidence of a scientific investigation for you to assume that the product is worth using?

2. Many research projects on new drugs give the medicine to half of the test participants but give a pill that does not contain the new medicine to the other half. Then the results for the two groups are compared. Give a reason why this is an important part of the experiment.

Name____________________________ Date__________ Class______________

Chapter 2 Standardized Test Preparation

1. Which of the following statements is the best summary of the rock cycle?

   A. Rocks deep below ground rise to the surface, are moved back underground, then rise to the surface again.

   B. Igneous rock and sedimentary rock change to metamorphic rock.

   C. The rock cycle has a single pathway from one type of rock to another type of rock.

   D. Every type of rock can be changed into every other type of rock. The type of rock that forms depends on the conditions that affect the rock.

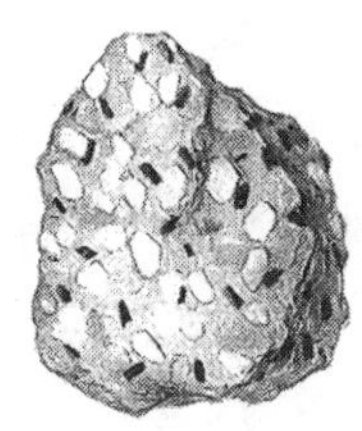
Syenite

Granite

Obsidian

Pegmatite

2. The pictures above show the textures of four igneous rocks that were collected during a field investigation. These rocks formed from magma that cooled at various rates. The faster magma cools, the smaller the crystals that form in the rock. Which of the above rocks formed from magma that cooled very rapidly?

   A. syenite C. obsidian

   B. granite D. pegmatite

3. Christine conduced a field investigation and discovered quartz is found in many different kinds of rock, including granite, gneiss, and sandstone. Based on this information, what could she conclude about quartz?

   A. Quartz is a mineral. C. Quartz is a sedimentary rock.

   B. Quartz is an igneous rock. D. Quartz is a metamorphic rock.

Name__________________________ Date__________ Class______________

Chapter 2 Standardized Test Preparation

4. The diagram shows the rock cycle. Analyze the diagram to determine which of the following steps in the rock cycle is necessary for sedimentary rock to form.

   A. heat and pressure

   B. melting and cooling

   C. uplift and erosion

   D. volcanic activity

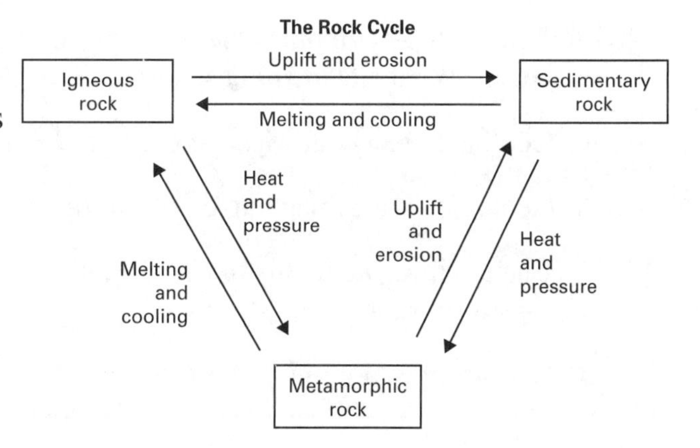

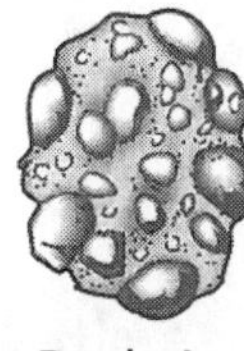
Rock A
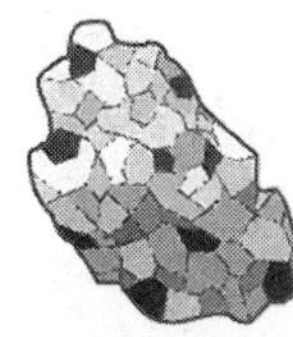
Rock B
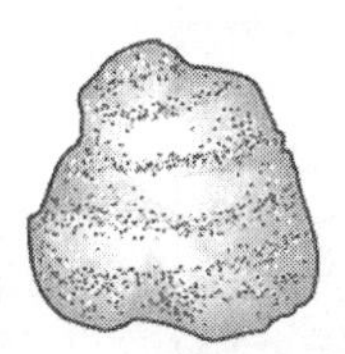
Rock C
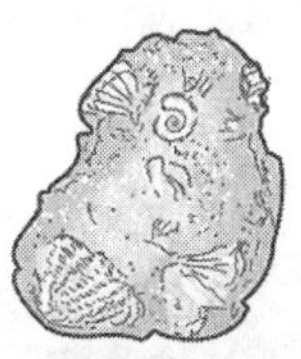
Rock D

5. Miguel went on a hike and found four different rocks. He wanted to identify what types of rocks he had found, so he collected data about them in the form of drawings and brief descriptions. According to Miguel's notes, Rock A would probably fall into which of the following categories?

   A. sedimentary C. metamorphic

   B. igneous D. mineral

6. What type of rock forms when heat and pressure change the structure, texture, or composition of sedimentary rock?

   A. igneous C. metamorphic

   B. sedimentary D. clastic

7. Which of the following is the most likely heat source for the formation of metamorphic rock?

   A. the heat from inside Earth

   B. the sun

   C. uplift

   D. the friction of movement of the plates against each other

Name______________________ Date__________ Class______________

Chapter 3 Standardized Test Preparation

1. Modern geology combines the ideas of uniformitarianism and catastrophism. What statement best summarizes this combination?

   A. Geologic change always happens gradually.

   B. Geologic change happened only in the past.

   C. Change in geologic history is gradual, but sometimes it is interrupted by sudden catastrophes.

   D. Catastrophes are much more common than gradual change.

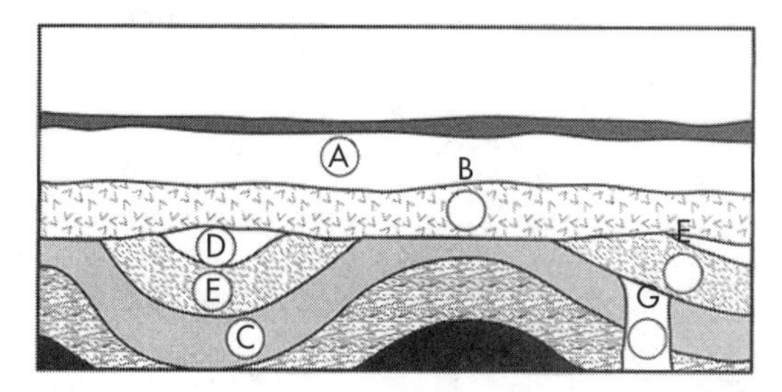

2. Look at the drawing above. Which principle would be used to determine the relative ages of the top three rock layers?

   A. catastrophism

   B. absolute dating

   C. superposition

   D. radioactive decay

3. During field investigations of undisturbed rock layers, scientists can assume that the relative age of a rock layer is probably

   A. less than the rock layer below it.

   B. more than the age of the fossils it contains.

   C. less than the age of the fossils it contains.

   D. determined by using radioactive decay.

4. George used radiometric dating in a laboratory investigation. He used this method because he wanted to find

   A. the relative ages of rocks.

   B. the absolute ages of rocks.

   C. the climate during a certain era.

   D. the types of fossils found in a rock.

Name______________________ Date__________ Class__________

Chapter 3 Standardized Test Preparation

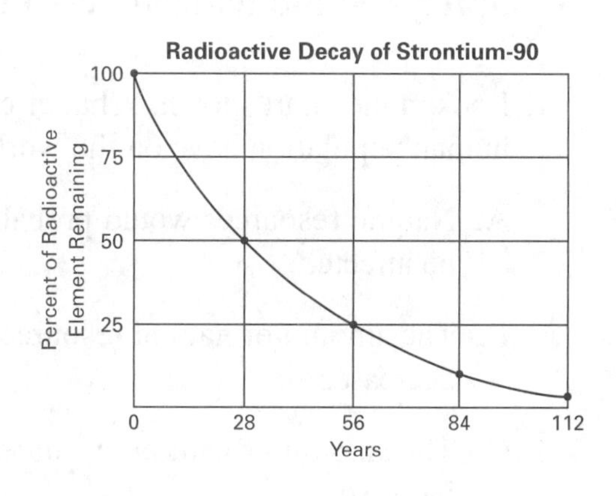

5. Ramona used this graph during an investigation about radioactive decay. The graph shows the radioactive decay rate of the isotope strontium-90. According to the graph, what is the half-life of strontium-90?

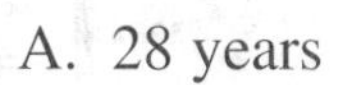

A. 28 years C. 84 years

B. 56 years D. 112 years

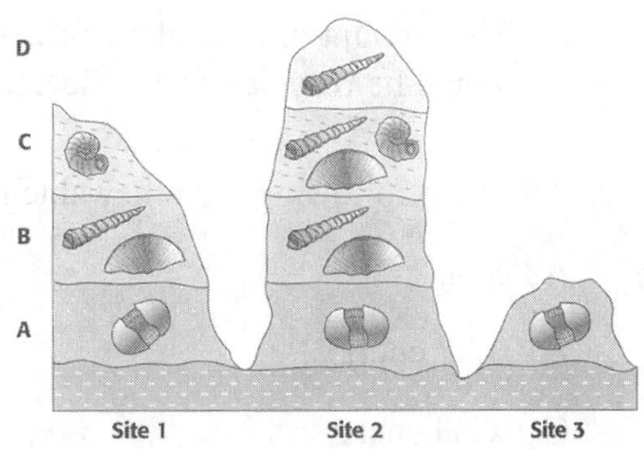

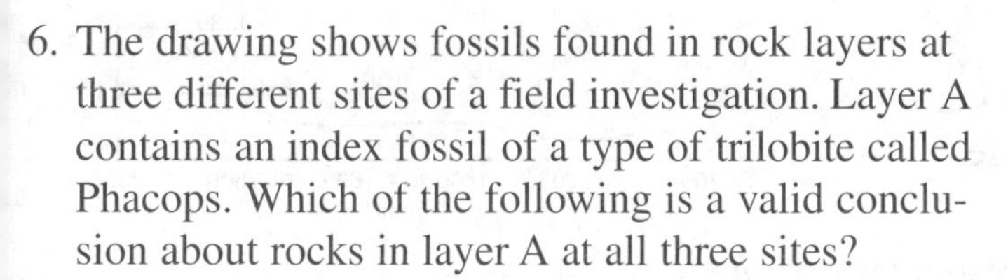

6. The drawing shows fossils found in rock layers at three different sites of a field investigation. Layer A contains an index fossil of a type of trilobite called Phacops. Which of the following is a valid conclusion about rocks in layer A at all three sites?

A. They were all formed at different times.

B. They are all different ages.

C. They are all about the same age.

D. They were all formed catastrophically.

The Geologic Time Scale

Era	Paleozoic							Mesozoic			Cenozoic	
Period	Cambrian	Ordovician	Silurian	Devonian	Mississippian	Pennsylvanian	Permian	Triassic	Jurassic	Cretaceous	Tertiary	Quaternary
Millions of years ago	540	490	443	417	354	323	290	248	206	144	85	76

7. The chart above shows part of the geologic time scale that was created from the studies of many geologists. According to the chart, what period was occurring 200 million years ago?

A. Tertiary C. Triassic

B. Cretaceous D. Jurassic

8. The largest divisions of geologic time are

A. eras C. periods

B. eons D. epochs

Name________________________ Date__________ Class______________

Chapter 4 Standardized Test Preparation

1. Look at the chart shown. What effects might a quickly growing human population have on the world's natural resources?

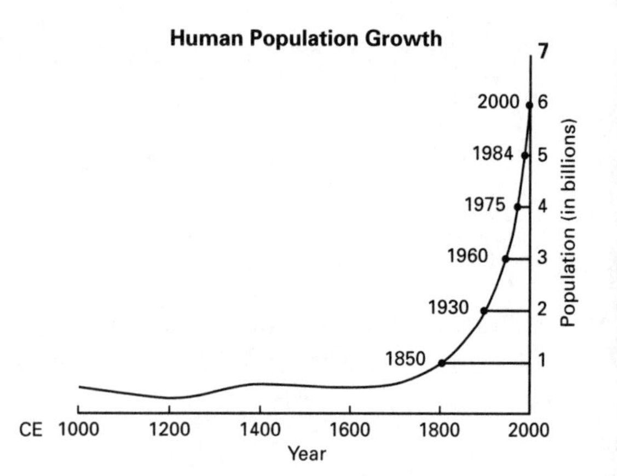

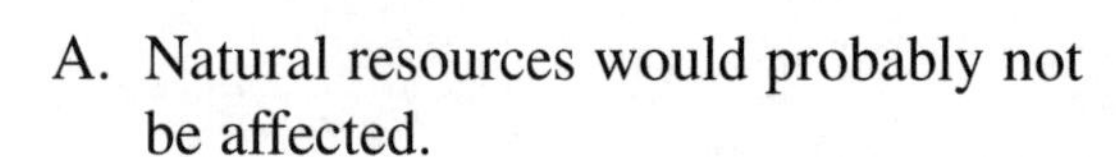

A. Natural resources would probably not be affected.

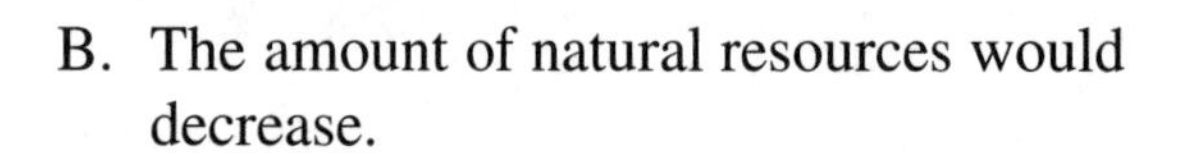

B. The amount of natural resources would decrease.

C. The amount of natural resources would increase.

D. The amount of natural resources that would be recycled would decrease.

2. All of the following are renewable resources *except*

A. coal.

B. solar energy.

C. wind energy.

D. geothermal energy.

3. A coal-burning power plant directly affects which of the following resources?

A. air

B. forests

C. water

D. sun

4. Which of the following statements about oil and natural gas reserves is **NOT** true?

A. Scientists find reserves by using seismic equipment.

B. Reserves are generally found under layers of impermeable rock.

C. Oil and natural gas are renewable resources.

D. Extracting reserves usually requires drilling through rock.

Name_______________________ Date_________ Class____________

Chapter 4 Standardized Test Preparation

5. What is one result of humans burning large amounts of fossil fuels?

   A. soil erosion

   B. abrasion

   C. contour plowing

   D. acid precipitation

6. If the supply of fossil fuels runs out, which of the following could potentially be used as an energy resource?

   A. nitrogen oxide

   B. water

   C. solar cells

   D. both B and C

7. Nuclear energy

   A. is renewable.

   B. produces smog.

   C. does not produce radioactive waste.

   D. is generated when the nucleus of a radioactive element splits.

8. Which of the following daily activities has the smallest effect on Earth's natural resources?

   A. turning off a light

   B. recycling old newspapers

   C. jogging through a park

   D. taking a bus to school

Name______________________________ Date__________ Class______________

Chapter 5 Standardized Test Preparation

1. Rushmi constructed the model of Earth shown. The labeled layers are all part of the

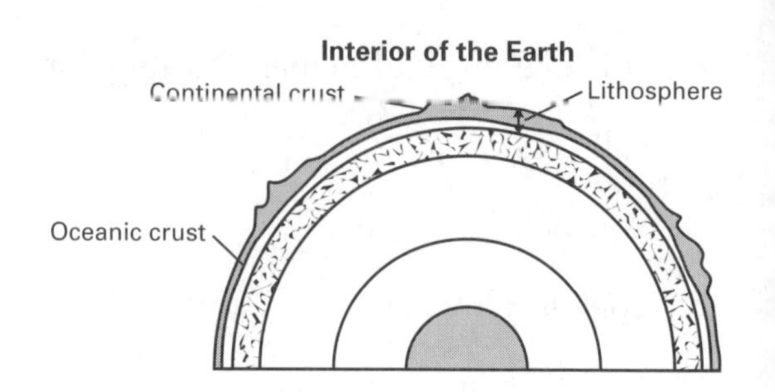

   A. outer core.

   B. inner core.

   C. crust.

   D. mantle.

2. Maria is about to conduct a field investigation about plate tectonics. She wants to study a region with transform boundaries. Look at the diagram. Which plate boundary would be a good choice for Maria to study?

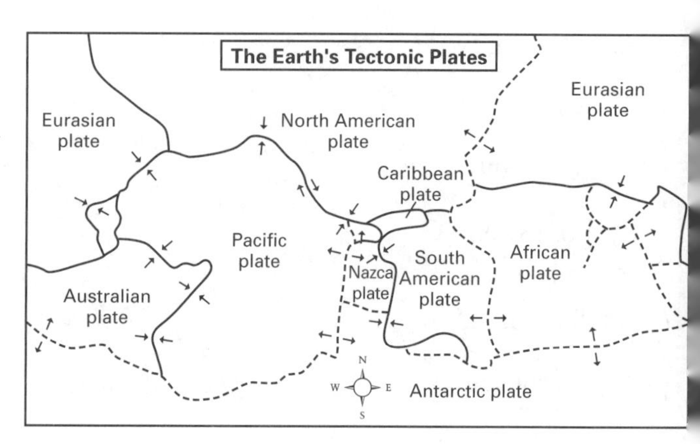

   A. Eurasian plate and North American plate

   B. Pacific plate and North American plate

   C. North American plate and Eurasian plate

   D. South American plate and Nazca plate

3. Which of the following is **NOT** explained by plate tectonics?

   A. earthquakes

   B. mountain building

   C. the temperature inside the Earth

   D. the supercontinent cycle

4. According to the theory of plate tectonics, what process occurs at a convergent boundary?

   A. Two tectonic plates push into each other.

   B. Two tectonic plates slide past each other.

   C. Two tectonic plates move away from each other.

   D. One tectonic plate moves up and over another.

Name________________________ Date________ Class____________

Chapter 5 Standardized Test Preparation

5. How is the rate of tectonic plate motion measured?

   A. by observing the change in mountain heights

   B. by surveying the distance between markers on different plates

   C. by analyzing data from Global Positioning Satellites

   D. It is so slow that it cannot be measured.

6. Geological evidence of seafloor spreading includes

   A. magnetic reversal.

   B. distribution of plant and animal species.

   C. the Rocky Mountains.

   D. ocean currents.

7. Reverse faults are usually caused by

   A. compression.

   B. tension.

   C. folding.

   D. a combination of compression and tension.

8. Which of these models best illustrates tectonic plate activity of the Earth's crust?

   A. the Earth as a solid rubber ball

   B. the Earth as an orange with the peel representing the crust

   C. the drawing in question 1 on the previous page

   D. ice cubes floating in a bowl of punch

Name____________________________ Date__________ Class______________

Chapter 6 Standardized Test Preparation

1. A large lava flow from a nonexplosive eruption is likely to cause

   A. damage to houses and roads.

   B. lower average global temperatures.

   C. a mudslide.

   D. buildings to collapse from volcanic ash.

2. Volcanoes erupt hot, liquid magma. What type of rock do volcanoes produce?

   A. metamorphic　　C. both metamorphic and igneous

   B. sedimentary　　D. igneous

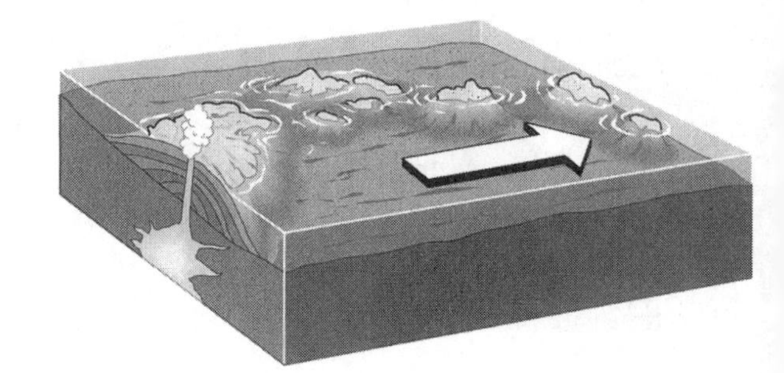

3. Examine the illustration depicting the formation of the Hawaiian Islands. One of the active volcanoes on the island of Hawaii is named Kilauea. If Kilauea is a shield volcano, which of the following conclusions is valid?

   A. Kilauea formed from repeated eruptions of low viscosity lava that have spread over a wide area.

   B. Kilauea formed from repeated moderately explosive eruptions of pyroclastic material.

   C. Kilauea is primarily made up of alternating layers of lava and pyroclastic material.

   D. Kilauea has formed a cinder cone through repeated eruptions of high viscosity lava.

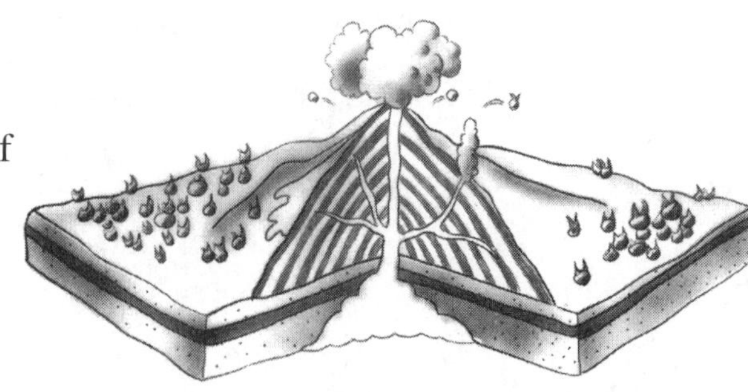

4. As part of a field investigation, Maureen made this sketch of a volcano. What type of volcano is shown in the sketch?

   A. dome　　C. shield

   B. composite　　D. hot spot

Name ____________________ Date ________ Class __________

Chapter 6 Standardized Test Preparation

5. On the map shown, note the two active volcanoes located within the Pacific Ocean that are not near a plate boundary. These volcanoes are most likely caused by

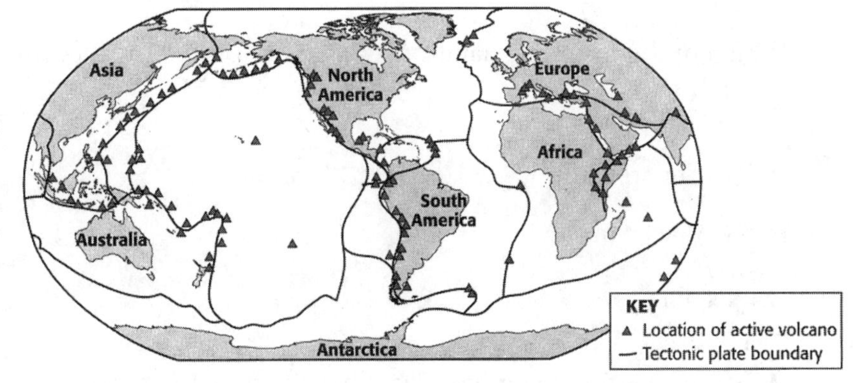

   A. hot spots.

   B. subduction zones.

   C. mid-ocean ridges.

   D. magma filling up gaps caused by a rift zone.

6. Based on the map above, which continent has the fewest active volcanoes?

   A. North America

   B. Africa

   C. Europe

   D. Australia

7. Based on the map above, which of the following is a valid conclusion?

   A. All volcanoes lie along tectonic plate boundaries.

   B. Hot spots are responsible for the active volcanoes in Africa.

   C. The majority of active volcanoes lie on plate boundaries surrounding the Pacific Ocean.

   D. There are only two volcanic hot spots in the world.

8. Which of the following is **NOT** used to predict volcanic eruptions?

   A. measurements of small earthquakes

   B. measurements of a volcano's slope

   C. measurements of volcanic gases

   D. measurements of a volcano's altitude

Name______________________ Date_________ Class____________

Chapter 7 Standardized Test Preparation

1. Which of the following landforms is **NOT** formed by deposition?

   A. sandbar

   B. barrier spit

   C. barrier island

   D. headland

2. Which of the following best describes the process responsible for the formation of sand?

   A. rock + waves = sand

   B. wind + waves = sand

   C. beach + longshore current = sand

   D. surf + wave train = sand

3. The coastal landform shown is known as a

   A. headland.

   B. sea stack.

   C. wave-cut terrace.

   D. sea arch.

4. Saltation is the

   A. skipping or bouncing of medium-sized (sand-sized) particles in the direction the wind is blowing.

   B. slow sliding or rolling of large particles (rocks) along the ground.

   C. process by which fine particles are carried through the air.

   D. process that combines all the erosion processes.

Name ____________________ Date __________ Class ____________

Chapter 7 Standardized Test Preparation

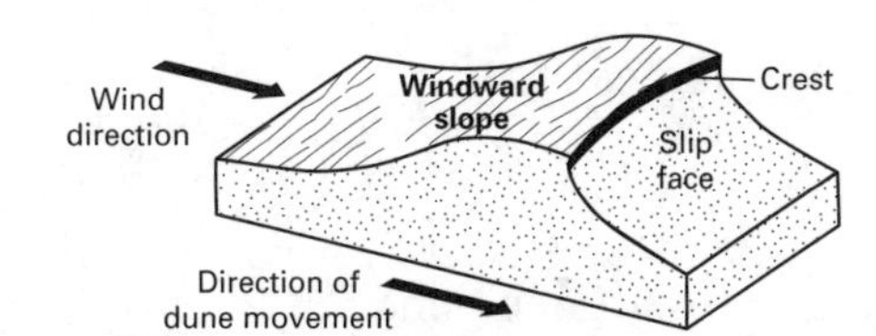

5. Tonya made the sketch above during a laboratory simulation of dune formation. The gently sloped side of the dune

   A. is called the slip face.

   B. slopes upward, perpendicular to the wind direction.

   C. faces the wind.

   D. slopes upward, perpendicular to the direction of the dune's movement.

6. Based on Tonya's sketch, which of the following is a valid conclusion?

   A. The type of dune shown in the sketch is known as desert pavement.

   B. Heavy rains encourage the formation of large dunes.

   C. A dune is a deposit of wind-blown particles.

   D. Longshore drift is occurring in the sketch.

7. Glaciers that form high in the mountains are known as

   A. alto glaciers. C. continental glaciers.

   B. alpine glaciers. D. moraines.

8. Human activities that can contribute to rock falls, landslides and mudflows include all of the following *except*

   A. road building.

   B. logging.

   C. heavy rains.

   D. house construction.

Name ____________________________ Date __________ Class ______________

Chapter 8 Standardized Test Preparation

1. The climate of Iceland is milder than that of Greenland because

   A. Greenland is farther north.

   B. the Gulf Stream carries warm water to Iceland but not Greenland.

   C. wind currents carry tropical air to Iceland but not Greenland.

   D. Iceland is closer to Europe.

2. Which biome is characterized by very low rainfall, frozen ground, and a large insect population?

   A. chaparral

   B. temperate desert

   C. tundra

   D. savanna

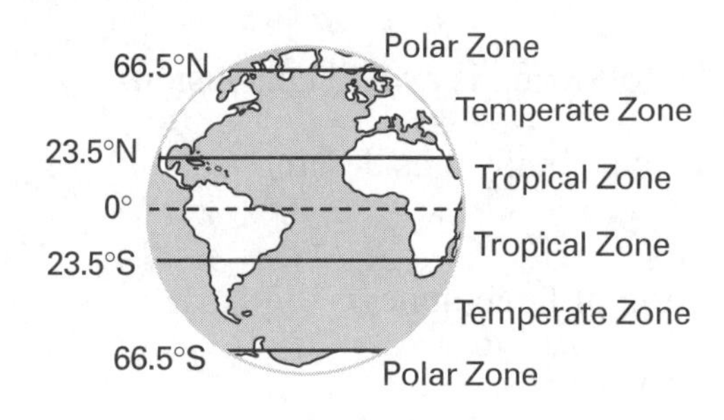

3. Look at the illustration of Earth's climate zones above. Biomes containing the largest number of plant and animal species are located

   A. in the polar zones.

   B. only in the northern portion of the tropical zones.

   C. in the tropical zones.

   D. in the temperate zones.

4. Deciduous trees lose their leaves in the winter. Which climate zone shown above contains forests of deciduous trees?

   A. the temperate zones

   B. the southern polar zone

   C. the northern polar zone

   D. the tropical zones

Name ____________________________ Date __________ Class ______________

Chapter 8 Standardized Test Preparation

5. The largest biome that is found near the equator is the

   A. taiga.

   B. tropical desert.

   C. temperate grassland.

   D. tropical rainforest.

6. Which of the following forces is responsible for breaking apart the supercontinent known as Pangaea?

   A. water pressure

   B. erosion

   C. air pressure

   D. tectonic forces

7. Which of the following natural events would most likely have an effect on the global climate?

   A. tornado

   B. volcanic eruption

   C. earthquake

   D. thunderstorm

8. According to scientists, which would be a direct result of global warming?

   A. heavy rain

   B. smog

   C. flooding

   D. greenhouse effect

Name____________________________ Date__________ Class______________

Chapter 9 Standardized Test Preparation

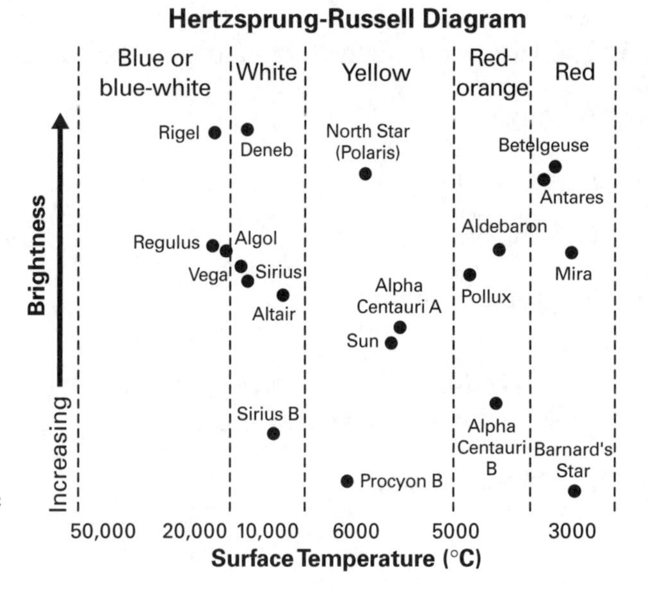

1. This diagram shows the Hertzsprung-Russell Diagram. According to the diagram, which of the following stars is both the dimmest and coolest?

   A. Rigel

   B. Betelgeuse

   C. Procyon B

   D. Barnard's star

2. Look at the diagram shown. Which of the following is **NOT** true about the star Rigel?

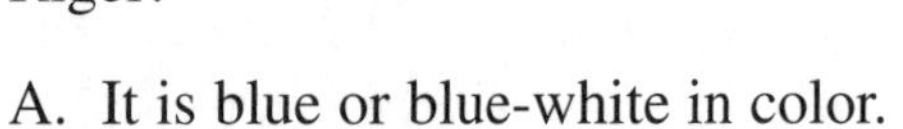

   A. It is blue or blue-white in color.

   B. It has a surface temperature of over 20,000°C.

   C. It is a dim star.

   D. It is a bright star.

3. Look at the diagram above. Which of the following is a star that is yellow and has an average brightness?

   A. Procyon B

   B. the sun

   C. Polaris

   D. Antares

4. Which of the following is **NOT** a characteristic of stars?

   A. Stars are made up of hot, dense gas.

   B. Stars are all the same size.

   C. Stars have apparent motion and actual motion.

   D. Stars can be different colors.

Name ______________________ Date __________ Class ____________

Chapter 9 Standardized Test Preparation

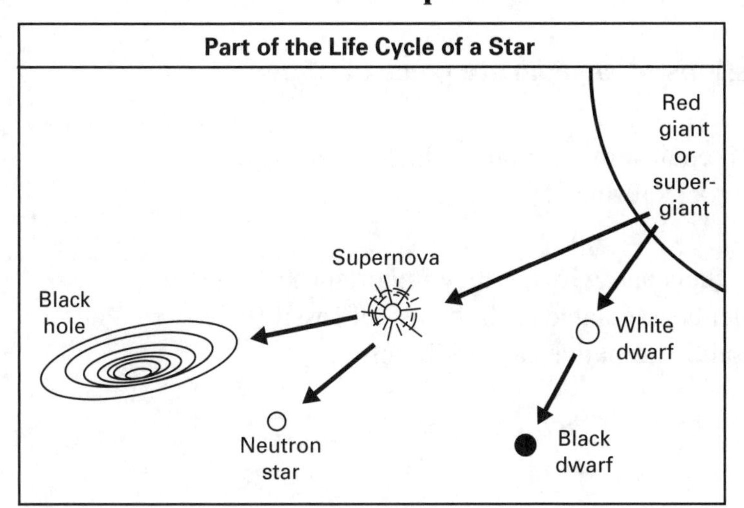

5. Look at the drawing above. Arrange the following according to the way they appear in the life cycle of a star.

   A. supergiant, black hole, supernova

   B. supergiant, neutron star, supernova

   C. supergiant, black dwarf, white dwarf

   D. red giant, supernova, black hole

6. Which of these is *least* likely to result from a supernova?

   A. galaxy

   B. neutron star

   C. pulsar

   D. black hole

7. Light travels from the sun to the Earth in about eight minutes, but from the next closest star it travels about four

   A. hours.

   B. days.

   C. months.

   D. years.

8. One piece of evidence that supports the big bang theory is the observation that most galaxies in the universe are moving

   A. away from each other.

   B. toward the Milky Way Galaxy.

   C. toward each other.

   D. toward the sun.

Name_________________________ Date_________ Class_____________

Unit 1 Standardized Test Preparation

Answer the following questions on a separate piece of paper.

1. Describe the conditions and environments under which igneous, sedimentary, and metamorphic rocks form.

2. What are the various fossil fuels and why are they important to humans? Discuss the problems associated with the use of fossil fuels, and the pros and cons of some alternatives to fossil fuels.

Name____________________________ Date__________ Class______________

Unit 1 Standardized Test Preparation

Answer the following questions on a separate piece of paper.

3. Write a few sentences to answer the following questions: Which is thicker, the continental crust or the oceanic crust? Why is the continental crust higher than the oceanic crust? Why do tectonic plates float and not sink into the Earth's mantle?

4. Describe some of the hazardous effects associated with earthquakes and volcanic eruptions and some technological advances that have helped humans predict and prepare for these events. Describe what people can do to help prevent injuries and loss of property in such events.

Name________________________ Date_________ Class____________

Unit 1 Standardized Test Preparation

Answer the following questions on a separate piece of paper.

5. How does our Sun compare to other stars? Explain what happens to stars like our Sun when they grow old and leave the main sequence portion of their life cycle. Then explain what happens to massive stars when they grow old and leave the main sequence. Other than size, in what ways do the most massive stars differ from low-mass stars?

6. Explain why soil monitoring and conservation are important and describe some of the consequences of failure to protect soil from erosion.

Name________________________ Date________ Class____________

Unit 1 Standardized Test Preparation

Answer the following questions on a separate piece of paper.

7. How might the technology of Global Positioning Systems (GPS) someday be used to predict earthquakes and other seismic events that are currently not predictable?

8. Explain how scientists combine radiometric dating and fossil records to establish the age of rock layers. How can this information be used to create and support theories about geological and biological evolution?

Name________________________ Date_________ Class____________

Chapter 10 Standardized Test Preparation

1. When two forces exerted on an object are balanced, the object

   A. speeds up.

   B. slows down.

   C. moves at constant velocity.

   D. does not experience a change in motion.

2. Why does a ball rolled across a carpet eventually come to rest?

   A. All motion must eventually come to a stop.

   B. The force of friction opposes the ball's motion.

   C. Balanced forces acting on the ball stop its motion.

   D. Equal and opposite forces are exerted on the ball.

3. Ball B is moving to the west. If a force toward the west acts on ball B,

   A. the ball would have negative acceleration.

   B. the velocity of the ball would decrease.

   C. the graph of the ball's motion would not change.

   D. the ball would travel a greater distance every second.

4. Which of the following forces acts against the upward motion of a rocket?

   A. rocket thrust

   B. gravity

   C. electrical forces

   D. magnetic forces

Name__________________________ Date__________ Class______________

Chapter 10 Standardized Test Preparation

5. When an object changes position over time when compared with a reference point, the object is

   A. accelerating.

   B. in motion.

   C. stopping.

   D. turning.

6. If you walk for 2 hours and travel 6 km, your average speed is

   A. 12 km/hr.

   B. 3 km/hr.

   C. 6 km.

   D. 2 km/hr.

7. A bus is moving north at 15 m/s, and you are walking to the rear of the bus at 1 m/s. Your resultant velocity is

   A. 14 m/s north.

   B. 16 m/s south.

   C. 1 m/s south.

   D. 15 m/s north.

8. Which of the following is **NOT** an example of acceleration?

   A. a car turning

   B. a bicycle speeding up as it goes down a hill

   C. a bus moving from rest to a speed of 40 km/hr

   D. a spaceship traveling at 400 km/hr

Name ____________________________ Date __________ Class ______________

Chapter 11 Standardized Test Preparation

1. One of Newton's laws states that an object traveling at a constant speed in a specific direction will continue to do so unless an unbalanced force acts on it. The moon orbits Earth because

   A. no unbalanced force acts on it.

   B. an unbalanced gravitational force constantly pulls the moon toward Earth.

   C. circular forces act on it.

   D. inertia pulls the moon toward Earth.

2. Newton's second law of motion explains why all objects fall with equal acceleration. Which of the following best summarizes this?

   A. Air resistance slows larger objects more than it slows smaller objects.

   B. Gravity exerts force on all objects, regardless of their size.

   C. The force-to-mass ratio is equal for all objects.

   D. Inertia is equal for all objects.

3. According to Newton's third law of motion, if an ice skater exerts a force on a wall, the

   A. wall exerts an equal and opposite force on the skater.

   B. acceleration of the wall depends on the magnitude of the force.

   C. wall will not move because of its inertia.

   D. momentum of the skater is no longer conserved.

4. A parachute slows the fall of a skydiver because of

   A. gravity.

   B. air resistance.

   C. acceleration.

   D. friction.

Name________________________ Date_________ Class____________

Chapter 11 Standardized Test Preparation

5. The orbit of the space shuttle around Earth is an example of

   A. constant motion.

   B. projectile motion.

   C. horizontal motion.

   D. a curved path.

6. Which of the following statements correctly describes the force that propels the space shuttle upward during takeoff?

   A. The gases push the shuttle upward with a force equal the shuttle pushing the gases downward.

   B. The gases push downward with a force equal to the shuttle pushing the gases downward.

   C. The shuttle pushes upward with more force than the gases push downward.

   D. The shuttle pushes downward with more force than the gases push upward.

7. What force is needed to accelerate a 1,500 kg car at a rate of 50 m/s/s?

   A. 1550 N

   B. 75,000 N

   C. 1450 N

   D. 30 N

8. A 100 kg football player crosses the goal line at a constant velocity of 6 m/s. His momentum is

   A. 94 kg/m/s.

   B. 16.6 kg/m/s.

   C. 106 kg/m/s.

   D. 600 kg/m/s

Name_______________________ Date_________ Class___________

Chapter 12 Standardized Test Preparation

1. A preservative added to a food product, such as juice, to slow down bacteria growth and prevent spoiling is an example of a(n)

   A. inhibitor.

   B. catalyst.

   C. exothermic reaction.

   D. endothermic reaction.

2. A compound that contains the name of a metal and a nonmetal is probably

   A. ionic.

   B. covalent.

   C. metallic.

   D. a liquid.

3. Which of the following is **NOT** a clue that a chemical reaction is occurring?

   A. gas formation

   B. precipitation of a solid

   C. change in the size and shape of a solid

   D. color change

4. One way that refrigeration preserves food is to slow chemical reactions by

   A. increasing the energy of the food.

   B. decreasing the frequency of collisions between molecules.

   C. causing the different substances in the food to separate.

   D. decreasing the activation energy.

Name______________________________ Date__________ Class______________

Chapter 12 Standardized Test Preparation

5. $$H_2CO_3 \rightarrow H_2O + CO_2$$
   The chemical reaction shown above is an example of a

   A. decomposition reaction.

   B. single-replacement reaction.

   C. synthesis reaction.

   D. double-replacement reaction.

6. What are the products of a double-replacement reaction if the reactants are NaCl and AgF?

   A. NaCl and AgCl

   B. NaF and AgCl

   C. NaF and NaF

   D. AgF and AgCl

7. In a chemical reaction, the chemical energy of the reactants is greater than that of the products. This is called a(n)

   A. endothermic reaction.

   B. spontaneous reaction.

   C. exothermic reaction.

   D. activation reaction.

8. Which of the following will **NOT,** in general, increase the rate of a reaction?

   A. increasing temperature

   B. increasing the concentration of reactants

   C. decreasing surface area

   D. adding a catalyst

Name__________________________ Date_________ Class____________

Chapter 13 Standardized Test Preparation

1. Which of the following is **NOT** an example of a covalent compound?

   A. water

   B. carbon dioxide

   C. sodium chloride

   D. sugar

2. When blue litmus paper is dipped in an unknown liquid, the litmus paper turns red. The liquid has a pH of

   A. less than 7.

   B. greater than 8.

   C. 7.

   D. greater than 12.

3. Which of the following is **NOT** a property of a base?

   A. tastes sour

   B. feels slippery

   C. changes the color of red litmus paper to blue

   D. will conduct electricity in a solution

4. Pure water has a pH of

   A. 4.

   B. 10.

   C. 8.

   D. 7.

Name____________________________ Date__________ Class______________

Chapter 13 Standardized Test Preparation

5. Which of the following chemical reactions does **NOT** produce a salt?

   A. neutralization of an acid and a base

   B. reaction of a metal with a base

   C. reaction of a metal and a nonmetal

   D. reaction of water and an acid

6. Which of the following is **NOT** a biochemical?

   A. carbohydrate

   B. lipid

   C. carbon dioxide

   D. protein

7. The largest molecules made by living things are

   A. enzymes.

   B. carbohydrates.

   C. lipids.

   D. nucleic acids.

8. Which of the following is **NOT** a function of proteins?

   A. regulating chemical activities

   B. providing energy for cellular activities

   C. transporting and storing materials

   D. providing structural support

Name____________________________ Date__________ Class______________

Chapter 14 Standardized Test Preparation

1. During a laboratory experiment about the nature of sound, a student walks into a large, dark room and yells "Hello!" She hears a strong echo of the word almost immediately. Which of the following is a valid conclusion that can be drawn from this observation?

   A. The room has smooth, hard walls and contains few objects.

   B. The room is full of pillows and other soft objects.

   C. The room has no walls.

   D. The room is very cold.

2. Which of the following is **NOT** a true statement about what happens when a guitar is played?

   A. The strings of the guitar are vibrating.

   B. Sound moves as a longitudinal wave away from the guitar.

   C. Particles of air are carried farther and farther away from the guitar with each wave.

   D. Sound moves away from the guitar in all directions.

3. Which of the following is **NOT** true of sound?

   A. It is caused by vibrations.

   B. It travels as a wave.

   C. It does not require a medium.

   D. It can travel through different media.

4. Where can sound **NOT** travel?

   A. through air

   B. through water

   C. through glass

   D. through empty space

Name________________________ Date__________ Class____________

Chapter 14 Standardized Test Preparation

5. Long-term exposure to loud sounds can cause tinnitus, a type of hearing loss, because loud sounds can damage the

   A. hair cells and nerve endings in the cochlea.

   B. eardrum.

   C. outer ear.

   D. anvil.

6. Sounds with a frequency above 20,000 Hz that can be used to "see" inside a patient's body are called

   A. infrasonic.

   B. supersonic.

   C. ultrasonic.

   D. X rays.

7. Hitting a drum harder increases the energy, or ____, of the sound it makes.

   A. frequency

   B. amplitude

   C. pitch

   D. length

8. A train whistle has a higher pitch when a train is approaching than it does when the train has passed. This is called

   A. the sound barrier.

   B. the Doppler effect.

   C. ultrasound.

   D. supersound.

Name_________________________ Date _________ Class ___________

Chapter 15 Standardized Test Preparation

1. Which of the following statements is true?

   A. Light is a form of electrical energy that travels through wires.

   B. Light is a form of energy that travels as a wave and can pass through different media.

   C. Light is generated by the sun and is captured for human use.

   D. Light is a process by which energy is generated.

2. During a medical technology experiment, a doctor observes images of broken bones. What type of wave is used to produce these images?

   A. X rays

   B. gamma rays

   C. visible light rays

   D. sound waves

3. In scientific notation, the speed of light in a vacuum, 300,000,000 m/s, is expressed as

   A. 3×10^{7} m/s.

   B. 300×10^{6} m/s.

   C. 3×10^{8} m/s.

   D. 3×10^{-8} m/s.

4. The electromagnetic wavelengths that can be perceived by the human eye are called

   A. the electromagnetic spectrum.

   B. colored light.

   C. visible light.

   D. the spectrum.

Name ____________________ Date __________ Class ____________

Chapter 15 Standardized Test Preparation

5. Which type of radiation from the sun causes wrinkles and skin cancer?

   A. infrared

   B. heat

   C. ultraviolet

   D. visible light

6. All types of electromagnetic waves have the same ____ in a vacuum.

   A. speed

   B. pitch

   C. wavelength

   D. amplitude

7. A black cat appears black because its fur ____ all the colors of visible light.

   A. reflects

   B. absorbs

   C. transmits

   D. refracts

8. Special cameras can be used to photograph animals at night because the animals' warm bodies give off more ____ rays than do their surroundings.

   A. ultraviolet

   B. light

   C. infrared

   D. ultrasonic

Name ________________________ Date __________ Class ____________

Unit 2 Standardized Test Preparation

Answer the following questions on a separate piece of paper.

1. At 20°C, sound travels at 343 m/s in air but at 5,200 m/s in steel. Why might a solid transmit sound faster than a gas? Is this true of all solids?

2. In a park on a bright summer day, sunlight reflects off many surfaces.

   a. Explain why a rose looks red.

   b. Why is it hotter in the sun than in the shade?

   c. Why does a calm pond look like a mirror, but grass doesn't?

Name______________________________ Date__________ Class______________

nit 2 Standardized Test Preparation

Answer the following questions on a separate piece of paper.

3. Explain what happens during a single-replacement chemical reaction and a double-replacement chemical reaction.

4. When a meteor enters Earth's atmosphere, it becomes hot enough to melt the metals in its structure. Describe the energy transfer that causes the meteor to become hot.

Name ______________________ Date ________ Class ____________

Unit 2 Standardized Test Preparation

Answer the following questions on a separate piece of paper.

5. Light and sound are both transmitted by waves. Describe how the two kinds of waves are similar and how they are different.

6. Explain how the technology of the wheel has changed the course of history by altering how people use force.

Name____________________________ Date__________ Class______________

Unit 2 Standardized Test Preparation

Answer the following questions on a separate piece of paper.

7. Explain what force keeps the International Space Station in orbit around the Earth.

8. Explain why you do not see a reaction force to motion of a ball falling toward Earth when you drop it, even though Newton's Third Law states that there is one.

Name____________________________ Date__________ Class______________

Chapter 16 Standardized Test Preparation

1. Which of the following is **NOT** a characteristic of all living things?
   A. They reproduce.
   B. They contain DNA.
   C. They make food.
   D. They grow and develop.

2. The compounds in cells that contain the information needed to make proteins are called
   A. lipids.
   B. nucleic acids.
   C. ATP.
   D. phospholipids.

3. An example of a producer is a
   A. bird.
   B. mushroom.
   C. salamander.
   D. pine tree.

4. Cells are surrounded by a cell membrane that is composed of
   A. lipids.
   B. phospholipids.
   C. nucleic acids.
   D. proteins.

Name______________________ Date__________ Class____________

Chapter 16 Standardized Test Preparation

5. Which of the following is **NOT** a benefit to an organism of being multicellular?

   A. larger size

   B. longer life

   C. specialization

   D. decreased energy needs

6. Which of these is something that all organisms require to live?

   A. oxygen

   B. carbon dioxide

   C. water

   D. sunlight

7. The basic forms of food that plant cells produce to store energy, and which are used by consumers, are

   A. lipids.

   B. carbohydrates.

   C. proteins.

   D. fats.

8. Organisms that are not producers require

   A. more space to live.

   B. specialized organs to digest food.

   C. more than one cell.

   D. a source of food.

Name____________________________ Date__________ Class______________

Chapter 17 Standardized Test Preparation

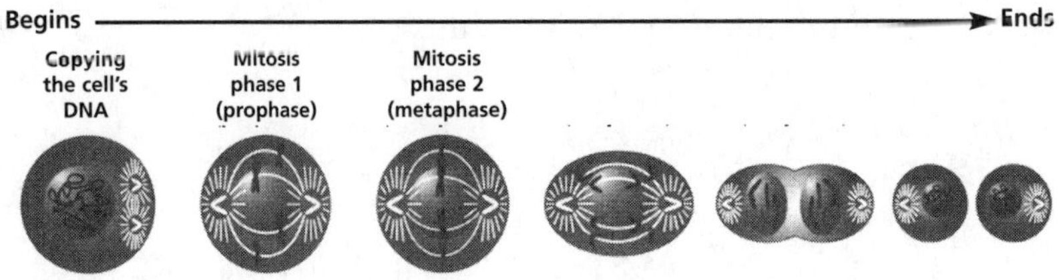

Begins → Ends

Copying the cell's DNA

Mitosis phase 1 (prophase)

Mitosis phase 2 (metaphase)

1. In the drawing of cell division shown, at which step do the chromatids line up along the equator of each cell?

   A. mitosis phase 1 (prophase)

   B. mitosis phase 2 (metaphase)

   C. mitosis phase 3 (anaphase)

   D. mitosis phase 4 (telephase)

2. Once mitosis is completed, the process of cytokinesis occurs. What happens during this process?

   A. The cytoplasm splits in two.

   B. The nuclear membrane breaks apart.

   C. The chromosomes line up.

   D. The chromatids separate.

3. Radiant energy from the sun is converted to chemical energy by the process of

   A. diffusion.

   B. osmosis.

   C. photosynthesis.

   D. mitosis.

4. A red blood cell placed in water

   A. will stay the same size because it has a cell membrane.

   B. will get smaller because of osmosis.

   C. will get bigger because of osmosis.

   D. will leak because of osmosis.

Name ____________________________ Date __________ Class ______________

Chapter 17 Standardized Test Preparation

5. Which of the following does **NOT** take place during active transport?

   A. Particles move from an area of low concentration to an area of high concentration.

   B. ATP is used.

   C. Particles pass through a membrane.

   D. Particles move from an area of high concentration to an area of low concentration.

6. Large particles are moved out of a cell by a process called

   A. endocytosis.

   B. exocytosis.

   C. active transport.

   D. diffusion.

7. Which of the following compounds is **NOT** released during cellular respiration?

   A. ATP

   B. CO_2

   C. H_2O

   D. O_2

8. Which of the following takes place when plant cells divide, but **NOT** when animal cells divide?

   A. A cell plate forms.

   B. The cell membrane pinches off.

   C. The cytoplasm splits.

   D. The nuclear membrane breaks apart.

Name ________________________ Date ________ Class ____________

Chapter 18 Standardized Test Preparation

1. A spherical bacteria resistant to drying is classified as a

   A. spirilla.

   B. bacilli.

   C. cocci.

   D. flagelli.

2. Which of the following is **NOT** a type of archaebacteria?

   A. methane maker

   B. heat lover

   C. salt lover

   D. oxygen maker

3. How are viruses like living things?

   A. They take in nutrients.

   B. They grow.

   C. They can live on their own.

   D. They can reproduce.

4. Which of the following is **NOT** a basic viral shape?

   A. crystal

   B. cylinder

   C. sphere

   D. spiral

Name__________________________ Date_________ Class_____________

Chapter 18 Standardized Test Preparation

5. Which of the following is **NOT** a bacterial disease?

   A. dental cavities

   B. food poisoning

   C. tuberculosis

   D. flu

6. The use of bacteria to clean up wastes and oil spills is called

   A. bioremediation.

   B. bioassay.

   C. biocleaning.

   D. nitrogen fixing.

7. Bacteria are **NOT** used to produce

   A. antibiotics.

   B. insulin.

   C. antiseptics.

   D. cheese.

8. Which of the following is the correct sequence of events in the lytic cycle?

   A. virus inserts genes into host cell, as host cell divides the virus genes are replicated

   B. virus enters host cell, virus takes over direction of host cell to produce more viruses, new viruses break out of host cell

   C. virus inserts genes into host cell, host cell directs replication of genes, genes break out of host cell

   D. virus enters host cell, host cell directs replication of more viruses, viruses break out of host cell

Name_________________________ Date_________ Class_____________

Chapter 19 Standardized Test Preparation

1. Which of the following is part of a salt marsh community?

   A. a river that empties into the marsh

   B. the soil

   C. the salt water that washes into the marsh

   D. hermit crabs

2. Which of the following is the correct order of the levels of environmental organization?

   A. organism, population, community, ecosystem, biosphere

   B. organism, community, population, ecosystem, biosphere

   C. organism, community, ecosystem, population, biosphere

   D. organism, population, ecosystem, community, biosphere

3. At each higher level of the food pyramid

   A. the amount of energy decreases.

   B. the number of organisms increases.

   C. the amount of energy increases.

   D. the size of organisms increases.

4. Which of the following organisms is a producer?

   A. mouse

   B. owl

   C. mushroom

   D. grass

Name ____________________ Date ________ Class __________

Chapter 19 Standardized Test Preparation

5. Which of these factors is not an essential part of determining the type of plants that can be grown in the controlled environment of a greenhouse?

   A. temperature
   C. size of the greenhouse
   B. humidity
   D. amount of light

6. A community of plants growing in a terrarium can grow even if not exposed to air outside the terrarium. Which of these possible explanations is NOT a reasonable hypothesis to explain how this can happen?

   A. There was enough carbon dioxide initially to support the growth for a long time.
   B. Some kind of decomposer that releases carbon dioxide is growing in the terrarium.
   C. The particular kind of plant in the terrarium does not require carbon dioxide.
   D. The sealed terrarium actually has small openings that allow exchange of air with the outside environment.

7. Which experiment below would NOT provide data that could support or disprove hypothesis B in question 6?

   A. Sterilize the soil before planting a second terrarium.
   B. Use a different kind of soil in a second terrarium.
   C. Examine the soil from the terrarium through a microscope to determine whether and microorganisms live in the soil.
   D. Add decomposers to the terrarium to determine whether any changes occur.

8. The terrestrial and aquatic food webs are connected to one another by which of these processes?

   A. erosion of rocks and minerals by streams
   B. leaves and other organic matter that decay in water
   C. the life cycle of phytoplankton
   D. benthic organisms

Name ______________________ Date ________ Class ____________

Chapter 20 Standardized Test Preparation

1. Which of the following would **NOT** be considered an example of pollution?

   A. chemicals

   B. noise

   C. heat

   D. oxygen

2. The release of CFCs into the atmosphere can cause which of the following?

   A. destruction of the ozone layer

   B. global warming

   C. poisoning of birds

   D. acid rain

3. The increase of carbon dioxide in the air would probably **NOT** cause

   A. an increase in global temperatures.

   B. the polar ice caps to melt.

   C. flooding of coastal areas.

   D. a decrease in sea level.

4. Which of the following actions would **NOT** help maintain biodiversity?

   A. protecting individual species

   B. protecting forests and wetlands

   C. preventing industrial wastes from being dumped in the ocean

   D. increasing consumption of natural resources

Name ____________________________ Date __________ Class ______________

Chapter 20 Standardized Test Preparation

5. Which of the following is a renewable resource?

   A. oil

   B. underground water

   C. coal

   D. solar energy

6. Which of the following strategies would **NOT** help conserve the environment?

   A. reduce pesticide use

   B. develop alternative energy sources

   C. protect habitats

   D. increase the number of landfills

7. Which of the following is **NOT** an example of the three Rs for conserving resources?

   A. reuse old clothes

   B. recycle plastics

   C. reduce the amount of electricity used

   D. refine the air

8. Which of the following statements about alien species is true?

   A. They are usually purposely introduced to their new habitat.

   B. They cannot drive out native species.

   C. They can become pests because they have no predators in their new habitat.

   D. They are not a problem because they cannot thrive in a new habitat.

Name____________________________ Date__________ Class____________

Chapter 21 Standardized Test Preparation

1. The correct path of food through the digestive tract is

   A. mouth, esophagus, stomach, liver, small intestine, rectum, anus.

   B. mouth, esophagus, stomach, small intestine, rectum, anus.

   C. mouth, esophagus, stomach, gallbladder, small intestine, rectum, anus.

   D. mouth, stomach, esophagus, small intestine, rectum, anus.

2. Which of the following is **NOT** a function of the liver?

   A. stores nutrients

   B. breaks down toxic substances in the blood

   C. makes bile to break down fats

   D. makes bicarbonate that neutralizes acids

3. The correct path of urine through the excretory system is

   A. kidneys, urethra, urinary bladder, ureter.

   B. kidneys, ureter, urinary bladder, urethra.

   C. urinary bladder, ureter, kidneys, urethra.

   D. ureter, kidneys, urinary bladder, urethra.

4. Antidiuretic hormone signals

   A. the kidneys to take back water from the nephrons and return it to the bloodstream.

   B. the kidneys to make more urine.

   C. the kidneys to take more water from the bloodstream.

   D. the nephrons to become more active.

Name__________________________ Date_________ Class____________

Chapter 21 Standardized Test Preparation

5. The main function of the urinary system is to

   A. break down toxic substances in the blood.

   B. remove waste products from the blood.

   C. remove solid wastes from the body.

   D. remove excess water from the body.

6. Which of the following is **NOT** a function of the stomach?

   A. mechanical digestion of food

   B. killing bacteria in food

   C. chemical digestion of food

   D. neutralizing the acid in chyme

7. An open sore in the stomach lining is called

   A. a gastric ulcer.

   B. stomach cancer.

   C. heartburn.

   D. a stomach tumor.

8. Nutrients are absorbed into the bloodstream in the

   A. stomach.

   B. liver.

   C. small intestine.

   D. large intestine.

Name__________________________ Date_________ Class____________

Chapter 22 Standardized Test Preparation

1. Which of the following diseases has been linked to exposure to chemicals in cigarettes?

   A. kidney disease

   B. autoimmune disorders

   C. heart disease

   D. skin cancer

2. Your body uses all of the following to fight viruses *except*

   A. killer T cells.

   B. helper T cells.

   C. B cells.

   D. V cells.

3. Rheumatoid arthritis is an example of a(n)

   A. allergic response.

   B. autoimmune disease.

   C. viral infection.

   D. bacterial infection.

4. A fever can be helpful in fighting an infection because it

   A. increases the rate of pathogen reproduction.

   B. increases the rate of T cell reproduction.

   C. causes memory cell to form.

   D. causes the blood to flow faster.

Name____________________________ Date__________ Class______________

Chapter 22 Standardized Test Preparation

5. The key concept of the germ theory proposed by Louis Pasteur is

   A. all organisms are made of cells.

   B. microorganisms cause disease.

   C. bacterial infections can be treated with antibiotics.

   D. illness results from an imbalance of body fluids.

6. Pasteur's initial experiments were designed to learn about

   A. animal diseases and how to prevent them.

   B. human diseases and how to prevent them.

   C. what causes milk and wine to spoil.

   D. the effect of bacteria in living systems.

7. Pasteur's work explained that cowpox can help prevent smallpox by

   A. helping the body build immunity.

   B. producing antibiotics that kill smallpox.

   C. creating a barrier to infection.

   D. competing with the smallpox infection.

8. Pasteur's discoveries led to changes in health practices that include all of the following *except*

   A. heating milk before it is sold in grocery stores.

   B. frequent handwashing.

   C. sterilization of surgical instruments.

   D. antibiotic treatments reduce the effects of allergies.

Name ______________________ Date ________ Class ____________

Chapter 23 Standardized Test Preparation

1. Which of the following is **NOT** caused by smoking tobacco?

   A. a decrease in heart rate and blood pressure

   B. an addiction to nicotine

   C. a loss of appetite

   D. an increase in the chances of developing lung cancer

2. What information is **NOT** included on a food label?

   A. serving size and number of servings per container

   B. a list of ingredients

   C. the percent of daily value (how much of each nutrient needed each day) contained in a serving

   D. the recommended daily servings from each food group

3. Which of the following food groups provide your cells with the essential building blocks they need to make proteins?

   A. meat, poultry, fish, dry beans, nuts and eggs

   B. fats and oils

   C. bread, cereal, rice and pasta

   D. fruits

4. Which of the following is the essential nutrient that is needed to transport substances, regulate body temperature, and provide lubrication?

   A. fats

   B. vitamins and minerals

   C. water

   D. carbohydrates

Name__________________________ Date __________ Class _____________

Chapter 23 Standardized Test Preparation

5. Which of these statements about natural chemicals is NOT true?

   A. They are made by natural processes.

   B. Because they are natural, they are safe.

   C. Some natural chemicals can be used as medicines.

   D. Water is one kind of natural chemical.

6. Medicines that are sold over-the-counter, without a doctor's prescription

   A. do not have any risks.

   B. can be taken as often as you like.

   C. can have harmful side effects if directions are not followed.

   D. are natural products, not chemicals.

7. Which of the following does NOT increase the amount of food your body needs to function well?

   A. exercise

   B. physical growth

   C. obesity

   D. being a boy compared to being a girl

8. You can catch a cold many times in one season because

   A. colds are caused by bacteria that pass from one person to another.

   B. there are many different viruses that cause cold symptoms.

   C. your body cannot make antibodies against cold viruses.

   D. colds are an autoimmune disorder.

Name______________________ Date________ Class__________

Unit 3 Standardized Test Preparation

Answer the following questions on a separate piece of paper.

1. Identify at *least* five ways in which all cells are alike.

2. Complex organisms have many levels of cell organization. Define each of the following levels: cells, tissues, organs, organ systems, organisms. Provide at *least* one plant and one animal example for each level, and describe how each level is related to the other levels.

Name__________________________ Date_________ Class____________

Unit 3 Standardized Test Preparation

Answer the following questions on a separate piece of paper.

3. What are producers, consumers, and decomposers, and how does each obtain the energy for life?

4. What is biodiversity? As part of your response, discuss ways in which biodiversity relates to the stability of an ecosystem. Describe at *least* two ways that human actions have reduced biodiversity, plus at *least* two actions that can be taken to help maintain or restore biodiversity.

Name________________________ Date_________ Class____________

Unit 3 Standardized Test Preparation

Answer the following questions on a separate piece of paper.

5. Explain what happens to an organism when its population grows too large for the resources available. Include a description of the types of limiting factors that affect population levels. How could you use the Internet and other computer resources to determine whether humans are at risk of overpopulating Earth?

6. Describe some of the problems that must be addressed if the human population of an area experiences a rapid population growth. Include a description of some kinds of resources that are needed to support a larger population and why their use must be carefully planned.

Name____________________________ Date__________ Class______________

Unit 3 Standardized Test Preparation

Answer the following questions on a separate piece of paper.

7. Describe the role of Memory B cells in your immune system. How do vaccinations use these cells to fight viruses?

8. Describe some of the techniques that have reduced the number of deaths by bacterial and viral infections over the past century.